Preface

This report presents findings from a research project conducted for the U.S. Army's Office of the Deputy Chief of Staff, G-3/5/7, Plans, Training, Mobilization, and Security Directorate (G-3), and the Army Capabilities Integration Center at Headquarters, U.S. Army Training and Doctrine Command, Fort Eustis, Virginia. The project began in the fall of 2011 under the dual sponsorship of these organizations.

The purpose of the study was to examine the evolving anti-access/area denial (A2AD) challenges that the U.S. military in general—and the Army in particular—would face in the 2012–2020 time frame.

The research effort involved developing a variety of plausible scenarios set in the 2020 time period in which joint and Army forces could be employed. Threats ranged from low- to high-intensity situations that would present a variety of challenges to the Army and the other services. These scenarios were used as the basis for Concept Options Group (COG) exercises (essentially, tabletop war games). These exercises led to important insights regarding A2AD threats to U.S. forces, as well as how those threats might be countered.

In addition, supplementary research examined how threats are evolving and how the United States might counter those growing threats.

This report will be followed by a restricted-distribution document that explains in detail the scenarios and the specific issues that were identified by each of the COGs and the related research. Meanwhile, this report serves to highlight the nature of the growing A2AD threats to U.S. forces and some possible counters to those threats.

This research was sponsored by the Army G-3 and the Army Capabilities Integration Center and conducted within RAND Arroyo Center's Force Development and Technology Program. RAND Arroyo Center, part of the RAND Corporation, is a federally funded research and development center sponsored by the United States Army.

The Project Unique Identification Code (PUIC) for the project that produced this document is RAN126154.

For more information on RAND Arroyo Center, contact the Director of Operations (telephone 310-393-0411, extension 6419; fax 310-451-6952; email Marcy_Agmon@rand.org), or visit Arroyo's web site at http://www.rand.org/ard.

Contents

RAND ARROYO CENTER

The Army's Role in Overcoming Anti-Access and Area Denial Challenges

John Gordon IV, John Matsumura

Prepared for the United States Army

The research described in this report was sponsored by the United States Army under Contract No. W74V8H-06-C-0001.

Library of Congress Cataloging-in-Publication Data

Gordon, John, 1956-
  The Army's role in overcoming anti-access and area denial challenges / John Gordon IV, John Matsumura.
    pages cm
  Includes bibliographical references.
  ISBN 978-0-8330-7993-0 (pbk. : alk. paper)
  1. Deployment (Strategy) 2. United States. Army. 3. Military doctrine—United States. 4. Operational art (Military science) 5. World politics—21st century. 6. Asymmetric warfare. I. Matsumura, John. II. Title.

  U163.G67 2013
  355.4—dc23

                                                    2013018910

The RAND Corporation is a nonprofit institution that helps improve policy and decisionmaking through research and analysis. RAND's publications do not necessarily reflect the opinions of its research clients and sponsors.

RAND˙ is a registered trademark.

Published 2013 by the RAND Corporation
1776 Main Street, P.O. Box 2138, Santa Monica, CA 90407-2138
1200 South Hayes Street, Arlington, VA 22202-5050
4570 Fifth Avenue, Suite 600, Pittsburgh, PA 15213-2665
RAND URL: http://www.rand.org/
To order RAND documents or to obtain additional information, contact
Distribution Services: Telephone: (310) 451-7002;
Fax: (310) 451-6915; Email: order@rand.org

Figures

Summary

The U.S. military has become increasingly concerned about the challenges it could face in gaining access to an operational area. Given their global responsibilities, the U.S. armed forces must be prepared to deploy to a wide range of locations that include almost any type of terrain and confront adversaries that span the threat spectrum from very poorly armed bands to peer-level foes. Research indicates that, in most situations, anti-access challenges require a joint solution, in which the capabilities of the different services can be brought to bear based on the threat and the mission. This study examined the nature of those future challenges and the Army's role as part of a larger joint or combined force.

Anti-access (A2) challenges prevent or degrade the ability to enter an operational area. These challenges can be geographic, military, or diplomatic. *Area denial* (AD) refers to threats to forces within the operational area. As they relate to U.S. ground forces (the Army and Marine Corps), AD threats are characterized by the opponent's ability to obstruct the actions of U.S. forces once they have deployed.

The Range of Anti-Access/Area Denial Threats

Irregular Adversaries

Irregular adversaries could be nonstate actors (such as a terrorist group or insurgents who have limited support from a state sponsor) or the armed forces of a very weak nation. It is likely that the A2 capability of such an opponent is limited.

Hybrid Adversaries

Hybrid adversaries have more and better weapons, operate with better command-and-control systems, are better trained, and are capable of simultaneously engaging in irregular and conventional combat. If a hybrid opponent is a nonstate actor, it is likely to have considerable support from a friendly nation or nations. Hezbollah in southern Lebanon in 2006 is the most notable recent example of this kind of opponent. For the purposes of this discussion, we assume that this class of threat includes both nonstate opponents that are capable of fighting in this manner and nation-states that elect to conduct operations in a hybrid mode.

State Adversaries

The U.S. military has historically planned for state adversaries, often assuming that lower-spectrum threats were lesser-included cases. State adversaries are the armed forces of another nation-state that normally operate in a conventional manner, wear uniforms, have formal military organizations, and employ the air, land, and naval weapon systems that are usually associated with traditional military forces. Their level of capability can vary enormously; such adversaries range from impoverished countries with obsolete military equipment to formidable potential opponents with large numbers of state-of-the-art military systems. Very few state opponents possess nuclear weapons, perhaps the ultimate A2 capability.

Key Anti-Access/Area Denial Capabilities

The following capabilities are most likely to pose significant anti-access/area denial (A2AD) challenges to joint operations in future scenarios:

- *Long-range precision-strike systems.* The development of Global Positioning System technology has increased the accuracy of both cruise and ballistic missiles. This increase in accuracy has profoundly changed the nature of the threat posed by long-range non-nuclear missiles, particularly against fixed targets (such as ports and airfields).
- *Littoral anti-ship capabilities.* These capabilities include high-quality non-nuclear submarines, fast missile-armed surface craft, and smart coastal and shallow-water mines.
- *High-quality air defenses.* During the famous Yom Kippur War of 1973, the state-of-the-art Soviet-made air defense system that challenged the Israeli Air Force was the SA-6 mobile surface-to-air missile (SAM) system. The SA-6 had a maximum effective range of roughly 25 km. Today, the state of the art in air defense is the Russian-built S-400 (SA-21) SAM. The S-400 has a maximum effective range against non-stealthy aircraft of approximately 400 km. Additionally, the low-altitude threat has become more challenging as improved anti-aircraft guns and shoulder-fired anti-aircraft weapons proliferate.
- *Long-range artillery and rocket systems.* The proliferation of mobile rocket launchers with ranges of more than 50 km can pose a major threat to Army or Marine Corps lodgment areas, such as ports and airfields.

A Joint Approach to Countering Anti-Access/Area Denial Threats

A review of U.S. operations since World War II, combined with the development and examination of possible future scenarios, indicated that—in most situations—the United States would employ a joint, interagency, intergovernmental, or multinational approach to overcoming diplomatic, geographic, and military A2AD challenges. The capabilities of the different services can be brought to bear in ways that are tailored for each specific operational environment.

There are important interdependencies and synergies between the services in terms of their ability to overcome A2AD challenges. For example, some threats to the Army's ability to deploy to an operational area must be addressed primarily by the other services (e.g., confronting the threat of enemy submarines and sea mines is the responsibility of the Navy, though the Army may be able to assist naval forces in mitigating those threats).

The Army has capabilities that can be of great value to the Air Force and Navy as they develop joint solutions to overcoming evolving A2 challenges. Specifically, the Army could provide considerable assistance to the Air Force and Navy in suppressing or destroying air defenses that challenge joint air operations. It could also facilitate dispersed operations by the

Air Force to assist the Navy in defeating littoral threats and establishing protected regional enclaves for naval operations.

As the services work together to develop operational concepts and systems to mitigate A2AD challenges, the Army will play a significant role in this effort.

Acknowledgments

We wish to express our gratitude to the members of our RAND Arroyo Center project team, who will be listed as coauthors on the longer restricted-distribution report that will follow this document: Matthew Boyer, David Johnson, David Frelinger, Thomas Herbert, Brian Nichiporuk, Todd Nichols, and Randall Steeb.

We also thank our Army sponsors, LTC Mike Whetstone and Elrin Hundley at Headquarters, U.S. Army Training and Doctrine Command, and MAJ Chris Williams of the Army G-3. At RAND, Jacob Heim provided important information on the Iranian and Chinese missile arsenals. We also thank CDR Phil Pournelle from the Office of Net Assessment in the Office of the Secretary of Defense, who provided important Navy-related insights. Hassan Rasheed and Michel Renninger of the National Ground Intelligence Center offered helpful input on Iranian regular and irregular forces. Finally, Robert Everson and RAND colleague Barry Wilson provided excellent reviews that enhanced the final version of the report.

Abbreviations

A2	anti-access
A2AD	anti-access/area denial
AD	area denial
ASBM	anti-ship ballistic missile
ATACMS	Army Tactical Missile System
C2	command and control
C4ISR	command, control, communications, computers, intelligence, surveillance, and reconnaissance
CEP	circular error probable
COG	Concept Options Group
FCS	Future Combat Systems
GPS	Global Positioning System
IED	improvised explosive device
ISR	intelligence, surveillance, and reconnaissance
JPADS	Joint Precision Air Drop System
MANPADS	man-portable air defense system
MRL	multiple-rocket launcher
NATO	North Atlantic Treaty Organization
SAM	surface-to-air missile
UAV	unmanned aerial vehicle

CHAPTER ONE

Background

The U.S. military has become increasingly concerned about the challenges it could face in gaining access to an operational area. Given their global responsibilities, the U.S. armed forces must be prepared to deploy to a wide range of locations that include almost any type of terrain and that span the threat spectrum from very poorly armed opposition to peer-level foes. Research indicates that, in most situations, anti-access challenges require a joint solution, in which the capabilities of the different services can be brought to bear based on the threat and the mission. This study examined the nature of those future challenges and the Army's role as part of a larger joint or combined force.

During the Cold War, much of the U.S. military—particularly the Army and Air Force—could plan on operating in regions where considerable forces were already deployed. For example, in 1988, as the Cold War was about to end, the U.S. Army had 207,000 personnel organized into two corps and the equivalent of five divisions in West Germany. At the same time, the Air Force had over 90,000 personnel and some 600 combat aircraft in Europe.[1] The anti-access challenge in that theater was characterized primarily by the enemy attempting to interfere with the arrival of reinforcements for the considerable forces that were already in the operational area prior to the start of hostilities.

Since the end of the Cold War, the U.S. military has become smaller and increasingly based inside the United States.[2] Starting with the 1991 Gulf War, the military has had to project power into regions where there has been little if any precrisis positioning of personnel and equipment. Potential adversaries have noticed this change in operational mode, and some are building capabilities to threaten the arrival and subsequent operations of U.S. forces as they attempt to deploy and initiate operations.

It is appropriate to start with a framework to ensure common understanding of the challenges that the Army, and the U.S. military in general, will likely face. It should be noted that while some of the terms are new, the U.S. military has confronted significant anti-access and area denial challenges in past operations.

Anti-access (A2) challenges prevent or degrade the ability to enter an operational area. These challenges can be geographic, military, or diplomatic.[3] For example, an operational area could be very far inland, a great distance from ports and usable airfields. That would be a geographic challenge. In other cases, diplomatic or political issues can pose an A2 challenge

---

[1] International Institute for Strategic Studies, *The Military Balance, 1988–1989*, London, 1988, p. 27.

[2] Even so, the number of locations to which the U.S. Army has been deployed since that time has been quite extensive.

[3] U.S. Army and U.S. Marine Corps, *Gaining and Maintaining Access: An Army–Marine Corps Concept*, version 1.0, March 2012, p. 3.

when one or more nations in a region prohibit or limit the ability of the U.S. military to deploy forces into their sovereign territory or to fly through their airspace.

Area denial (AD) refers to threats to forces within the operational area. As they relate to U.S. ground forces (the Army and Marine Corps), AD threats are characterized by the opponent's ability to obstruct the actions of U.S. forces once they have deployed. Importantly, there are far more potential opponents that could pose significant AD challenges than there are opponents with major A2 capabilities. For example, when U.S. forces deployed to Afghanistan in 2001–2002, there was not a significant military A2 threat, although there were initially diplomatic challenges to overcome with regard to nearby countries, and the geography of the region required a long-distance deployment far from the sea and existing U.S. bases. However, once U.S. forces began operating in Afghanistan, they faced numerous and, at times, severe AD threats, such as the increasingly common use of improvised explosive devices (IEDs) that caused casualties and imposed constraints on the mobility of U.S. and coalition forces.

The types of anti-access/area denial (A2AD) threats that the U.S. military could encounter in future operations will vary considerably. At the *low end* of the conflict spectrum, there could be guerrilla-type forces, like the Taliban in Afghanistan, with very limited A2 capabilities and a small number of modern weapons. These forces could still pose a considerable AD challenge due to their ability to operate among the local population and employ irregular tactics to strike U.S. forces at times and places of their choosing.

In the *middle* of the spectrum are so-called "hybrid" opponents, which can employ irregular or guerrilla-type tactics but are reasonably well armed with modern weapons. Hybrid opponents can therefore simultaneously fight in a conventional manner. Examples include the irregular Viet Cong and regular North Vietnamese forces during the Vietnam War and, more recently, the Hezbollah forces that Israel fought in southern Lebanon in 2006.[4]

At the *high end* of the threat spectrum are the armed forces of nation-states that tend to employ conventional tactics and weapons. Even at this end of the spectrum, the level of A2AD capability can vary considerably. As with the hybrid threat, this challenge is not new to the U.S. military. In the case of World War II, Nazi Germany had a potent, long-range A2 capability in its submarine force (the U-boats) that threatened Allied shipping routes that carried troops and supplies across the Atlantic. Similarly, during the Cold War, a major mission of the Soviet Navy's submarines was to prepare to interdict the movement of U.S. reinforcements to Europe.

In many cases, the U.S. military will have to employ a system of joint capabilities to overcome A2AD challenges. This observation is based on both the insights gained in the scenarios that were examined as part of this research and an examination of how operations were actually conducted in the post–World War II era in which a range of air, land, and naval capabilities were required to gain and maintain access. In some situations, U.S. air and naval power will be the primary capabilities required (at least in the initial phases of an operation) to overcome significant A2 threats. In other situations, the role of ground forces will dominate or could come to do so as an operation progresses.

Chapter Two of this report examines the range of A2AD military threats that U.S. forces could confront today or in the foreseeable future, including the relationship of air, land, and naval forces in overcoming those threats. Chapter Three explores a selection of key threat

---

[4]   Frank G. Hoffman, "Hybrid Warfare and Challenges," *Joint Force Quarterly*, No. 52, 1st Quarter, 2009.

capabilities in greater detail, with an emphasis on the joint implications of A2 challenges. Chapter Four profiles a joint approach to countering A2AD challenges, focusing specifically on the A2 challenges faced by the Air Force, the Navy, and the Army and options to reduce these threats. Chapter Five offers conclusions and highlights the primary findings presented in the report.

The Range of Anti-Access/Area Denial Threats

In this chapter, we review the general types of A2AD threats that U.S. forces face today and will likely confront in the future. Some of the A2 threats presented here are of particular concern and are discussed in greater detail in the next chapter.

Irregular Adversaries

Irregular adversaries could be nonstate actors (such as a terrorist group or insurgents who have limited support from a state sponsor) or the armed forces of a very weak nation. It is likely that the A2 capability of such an opponent is limited.[1]

Anti-Access Threats

A2 threats that U.S. forces could encounter include the following:

- small, high-speed boats used to attack shipping vessels in coastal regions, either with onboard weapons or via suicide ramming[2]
- mortars capable of shelling airfields or ports from ranges of up to roughly 10 km
- artillery rockets, launched either singly or in small numbers, capable of striking targets up to roughly 20 km away.[3]

Area Denial Threats

AD threats could include the following:

- sniper attacks, which may be effective in slowing the flow of aircraft into and out of airports, or more-coordinated attacks involving automatic weapons, such as heavy machine guns capable of threatening low-altitude aircraft as they land or take off from airports
- IEDs and conventional land mines of various types, which have been very effective, low-tech, and low-cost insurgent weapons in Iraq and Afghanistan and have compelled the United States to take major steps to reduce the threat

---

[1] David E. Johnson, *Hard Fighting: Israel in Lebanon and Gaza*, Santa Monica, Calif.: RAND Corporation, MG-1085-A/AF, 2011, pp. 148–151.

[2] "USS *Cole* Bombing: Suicide Attack," *GlobalSecurity.org*, last updated October 11, 2006.

[3] An example would be the Soviet-era 122-mm unguided artillery rockets that are commonly available around the world today.

- individuals or small units operating with light infantry weapons, such as small arms or machine guns, likely in organized groups no larger than a company (roughly 100–200 personnel)
- individuals or units with man-portable anti-armor weapons, such as rocket-propelled grenades[4]
- manned or light vehicle-portable mortars, probably without guided projectiles
- simple electronic warfare attacks, such as low-tech jamming of Global Positioning System (GPS) signals.[5]

Hybrid Adversaries

Compared with irregular forces, hybrid adversaries have more and better weapons, operate with better command-and-control (C2) systems, are better trained, and are capable of simultaneously engaging in irregular and conventional combat.[6] If a hybrid opponent is a nonstate actor, it is likely to have considerable support from a friendly nation or nations. Hezbollah in southern Lebanon in 2006 is the most notable recent example of this kind of opponent. In Hezbollah's case, there was considerable support from both Iran and Syria, including the provision of long-range rockets and anti-ship cruise missiles. There are historical examples of nations fighting in this manner as well. During World War II, German forces in the Soviet Union were simultaneously confronted with a sizable state-controlled guerrilla threat in their rear area (the "Partisans") and engaged in a massive conventional battle against a peer-level opponent in the form of the Soviet Army and Air Force. The mix of irregular and conventional tactics employed by the Viet Cong and the North Vietnamese regular military is another example of a nation-state operating in this manner.

Another important distinction of this class of threat is that the opponent is able to operate in larger formations and control terrain, at least temporarily.[7] For the purposes of this discussion, we assume that this class of threat includes both nonstate opponents that are capable of fighting in this manner and nation-states that elect to conduct operations in a hybrid mode. In the latter case, U.S. forces could simultaneously be confronted by a hybrid threat and strictly conventional capabilities, discussed in the section "State Adversaries," later in this chapter.

Although the exact nature of the military A2 threat would vary depending on the opponent, examples of the opponent's capabilities could include the threats listed in the case of irregular adversaries, plus the following:

- C2 capabilities to plan and conduct operations with up to roughly a battalion-size force
- short-range anti-ship weapons, such as guns and missiles, that can be used within line-of-sight distances
- small numbers of shallow-water sea mines

---

[4] It should be noted that small arms, machine guns, rocket-propelled grenade launchers, and similar weapons can also be used against low-flying aircraft or helicopters.

[5] Depending on the capability of the irregular opponent, this could be an A2 challenge.

[6] Johnson, 2011, pp. 151–162.

[7] Johnson, 2011, pp. 151–163.

- cannons, mortars, and rocket launchers capable of operating in battery- or battalion-size firing elements at ranges up to 100 km with precision warheads, thus presenting a significant threat to airfields and ports (see Figure 2.1)[8]
- limited numbers of unmanned aerial systems for both surveillance and attack, adapted commercially available systems.

In this case, AD threats could include those listed in the irregular adversary case, plus the following:

- man-portable air defense systems (MANPADS) able to threaten aircraft that lack appropriate countermeasures while flying at low altitude
- 37- to 57-mm anti-aircraft guns capable of threatening aircraft flying at up to 10,000 ft
- short-range precision weapons, such as anti-tank guided missiles equipped with modern warheads that can attack at ranges of 2–5 km
- small numbers of older, armored vehicles

Figure 2.1
Hezbollah's Rocket and Missile Threat to Israel, 2006

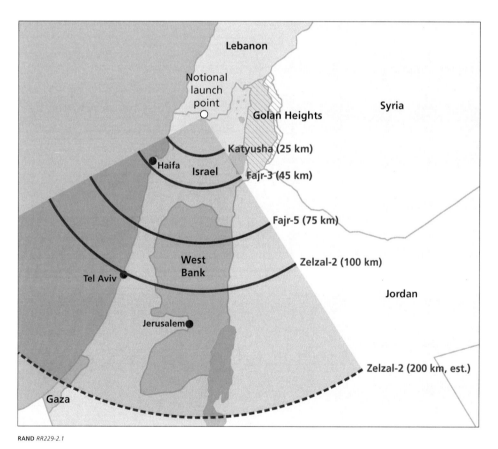

---

[8] Examples include the variety of rockets and missiles available to Hezbollah during the 2006 conflict in southern Lebanon.

- limited cyber attack capabilities intended to disrupt and degrade U.S. and coalition command, control, communications, computers, intelligence, surveillance, and reconnaissance (C4ISR) systems
- chemical or biological weapons.[9]

State Adversaries

To a large extent, the U.S. military has historically planned for state adversaries, often assuming that lower-spectrum threats were lesser-included cases.[10] This tendency was perhaps most prevalent in the U.S. Army and Marine Corps. Interestingly, during the past decade's operations in Afghanistan and Iraq, the Army and Marine Corps had to reorient toward irregular and hybrid-type opponents to a considerable extent. During that period, the Air Force and Navy, generally in a supporting role to the Army and Marine Corps, remained more focused on state-to-state conflict. As the wars in Iraq and Afghanistan waned, the Air Force and Navy started to highlight the A2 (and, to a lesser extent, AD) potential of possible future opponents.

State adversaries are the armed forces of another nation-state that normally operate in a conventional manner, wear uniforms, have formal military organizations, and employ the air, land, and naval weapon systems that are usually associated with traditional military forces. Their levels of military capability can vary enormously; such adversaries can range from impoverished countries with obsolete military equipment to formidable potential opponents with large numbers of state-of-the-art military systems. We address nuclear weapons as a unique military capability later in this report.

The following are some of the more significant military A2 capabilities that better-armed nations possess or could obtain that would be particularly threatening to U.S. forces:

- naval forces, including both surface ships and attack submarines, able to attack U.S. warships and commercial shipping vessels up to thousands of miles offshore
- long-range anti-ship ballistic missiles (ASBMs) and cruise missiles[11]
- naval mines, including sophisticated bottom mines that can sit quietly on the seabed, waiting for the most valuable ship target to be detected by their onboard sensors
- long-range precision, conventionally armed cruise and ballistic missiles for strikes against ports and airfields; long-range strike aircraft capable of attacking ships, ports, air bases, C2 facilities, and deployed forces ashore; and weapons aboard the aircraft, which could include a mix of gravity weapons (guided or unguided) and long-range precision stand-off munitions, such as air-launched cruise missiles

---

[9]   Even nonstate hybrid opponents may be able to develop and field these types of weapons; they could be employed in traditional ways (e.g., fired in artillery rounds) or clandestinely (e.g., smuggled into a location in inconspicuous containers).

[10]   Andrew F. Krepinevich, Jr., *The Army and Vietnam*, Baltimore, Md.: Johns Hopkins University Press, 1988, pp. 4–7.

[11]   Although anti-ship cruise missiles have been available for decades, the emergence of the ASBM is very recent and so far limited to the Chinese armed forces. See "Chinese Develop Special 'Kill Weapon' to Destroy U.S. Aircraft Carriers," U.S. Naval Institute, March 31, 2009.

- unmanned aerial systems for strike, surveillance, and reconnaissance, the ranges and pay-loads of these which are steadily increasing[12]
- long-range surface-to-surface rockets with ranges of more than 50 km. Warheads include unitary and submunitions, and can be precision-guided
- long-range air defenses capable of engaging non-stealthy aircraft at ranges of more than 100 km, often resistant to many electronic counter-measures[13]
- advanced low-to-medium altitude air defense systems (including advanced MANPADS and radar/infrared anti-aircraft guns)
- electronic and cyber warfare techniques and technologies targeting communication networks, GPS, and air- and space-based reconnaissance and surveillance capabilities.

The types of AD threats that this class of opponent could field include the following:

- armored fighting vehicles, perhaps in large numbers
- attack helicopters and fixed-wing strike aircraft, possibly armed with precision munitions
- large numbers of cannons, heavy mortars, and rocket launchers, including precision munitions
- long-range, precision anti-armor systems, including both anti-tank guided missiles and indirect-fire and artillery-delivered weapons
- possibly significant chemical or biological weapon stocks.[14]

Chapter Three explores the more significant threats that U.S. forces might encounter in future operations.

---

[12] A number of nations have fielded classes of systems with sensors that can be used for real-time reconnaissance and surveillance, as well as systems that can conduct attacks either by launching weapons that they carry or by crashing into their targets. See "China Building an Army of Unmanned Military Drones 'to Rival the U.S.,'" *Daily Mail*, July 5, 2011.

[13] Examples include the Russian and Chinese S-300/400 system. See Carlo Kopp, "Almaz S-300—China's 'Offensive' Air Defense," International Assessment and Strategy Center, February 25, 2006.

[14] Such weapons would probably be employed through traditional military means, such as via multiple-rocket launchers (MRLs) that can quickly deposit fairly significant amounts of agent into a target area. That said, a state adversary could also elect to deliver chemical or biological weapons via clandestine means.

Key Threat Capabilities

Chapter Two provided an overview of the types of military A2AD threats that the U.S. military currently faces and is likely to confront in future operations. As noted in that chapter, the severity of threats varies considerably, depending on the capability of the opponent. In some cases, a nonstate opponent, such as Hezbollah, might be better armed than the military forces of a weak nation-state.

This chapter examines some particularly challenging threats in greater detail. Many of the threats examined in this chapter have major implications for the Air Force and Navy. This matters to the Army, because if enemy A2 threats force the Navy and Air Force to change their mode of operations, that could have a significant effect on how the Army employs its forces. Keeping with that theme, this chapter focuses primarily on new or emerging military A2 challenges. We conclude with a brief discussion of a particularly serious A2 challenge: nuclear-armed opponents.

The Rise of Long-Range Precision Strike Systems

The development of GPS technology has enabled a great increase in the accuracy of both cruise and ballistic missiles. This increase in accuracy has profoundly changed the nature of the threat posed by long-range non-nuclear missiles, particularly against fixed targets, such as ports and airfields.

Ballistic Missiles

The increase in the number and variety of precision ballistic missiles is a particularly troubling threat—and one that is very different from the inaccurate Scud-class weapons that made sensational, but militarily ineffective, attacks on U.S. and coalition military forces and civilian targets during the 1991 Gulf War. Whereas the 1960s-era Scud-B had a circular error probable (CEP, within which 50 percent of missiles will land) of some 900 m when fired at a target 300 km distant,[1] the modern Chinese DF-15A missile has a CEP of 30–45 m, and the DF-15B has a CEP of 5–10 m, at a maximum range of 600 km (see Figure 3.1).[2] This type of missile capability presents a significant threat to ports and, especially, airfields. Warheads can be either unitary (with a single high-explosive charge) or scatterable submunitions.

---

[1] "R-11/SS-1B SCUD-A, R-300 9K72 Elbrus/SS-1C SCUD-B," Federation of American Scientists, last updated September 9, 2000.

[2] "DF-15/-15A/-15B (CSS-6/M-9)," *Missile Threat*, last updated January 5, 2013.

Figure 3.1
Chinese DF-15 Missile

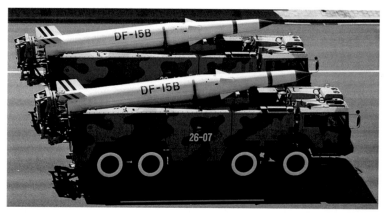

RAND *RR229-3.1*

It is important to note that while the number of nations with missile arsenals is growing, the quality and number of these weapons varies considerably. China and Russia, for example, are very well equipped, with a large number of long-range precision weapons. Iran and North Korea also have hundreds of missiles, but their arsenals are composed primarily of older weapons with poor accuracy. Figure 3.2 shows the difference in the number of Chinese and Iranian theater ballistic missiles and ground-launched cruise missiles.

In addition to the considerable difference in the number of missiles, the level of threat posed depends to a great degree on the accuracy of these weapons. For example, if an opponent were attempting to shut down an air base with missile strikes, the number of weapons that it would have to fire would be highly dependent on the missiles' accuracy. When comparing China and Iran, not only is the Chinese missile arsenal considerably larger, as shown in Figure 3.2, but its weapons are also much more accurate. Figure 3.3 shows the number of missiles that Iran would have to fire to have a high probability of hitting a target 100 m in diameter, such as an aircraft hangar or a group of large fuel tanks on an air base.

Guided cruise and ballistic missiles tend to be fairly expensive and, thus, may be available in only limited numbers. Additionally, for nations that possess these kinds of advanced capabilities, the norm is for the larger portion of the missile inventory to consist of shorter-range weapons (which are cheaper and can be produced in greater numbers), with fewer of the more expensive, longer-range systems. Unfortunately, these kinds of capabilities are proliferating at a relatively high rate. Part of the reason for this is the reduction in the cost of the underlying technology. Another reason is the cost-effectiveness of this solution relative to others (e.g., developing and maintaining a large, modern air force).

Cruise Missiles and Unmanned Aerial Vehicles

Ballistic missiles have the advantage of great range and speed of flight. They are expensive systems, however. Cruise missiles and unmanned aerial systems (which can be armed or unarmed and are also used for reconnaissance and surveillance) are much cheaper and do not require as advanced a technical base as do ballistic missiles. The proliferation of unmanned aerial vehicles (UAVs), in particular, has been significant over the past decade. Hostile reconnaissance UAVs can monitor and report on the activities of U.S. Army forces in an operational area and provide

Figure 3.2
Chinese and Iranian Ballistic and Cruise Missile Inventories, by Range

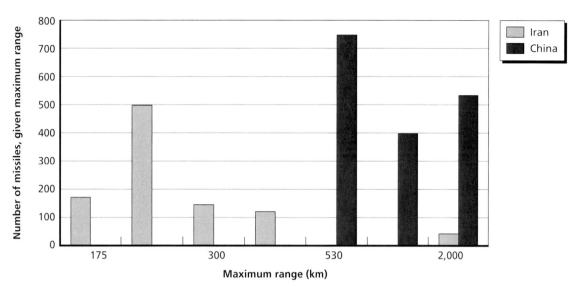

SOURCES: Ballistic and cruise missile range data from *Jane's Strategic Weapons Systems*; Office of the Secretary of Defense, *Annual Report to Congress: Military and Security Developments Involving the People's Republic of China*, Washington, D.C., 2011; and National Air and Space Intelligence Center, *Ballistic and Cruise Missile Threat*, Wright-Patterson AFB, Ohio, NASIC-1031-0985-09, 2009. Missile inventory data from International Institute for Strategic Studies, *The Military Balance 2010*, London, February 2010; Office of the Secretary of Defense, 2011; and Office of the Secretary of Defense, *Unclassified Report on Military Power of Iran*, Washington, D.C., April 2010.
NOTE: The figure includes upper estimates for both range and inventory.
RAND *RR229-3.2*

Figure 3.3
Number of Chinese and Iranian Missiles Required Against Point Targets

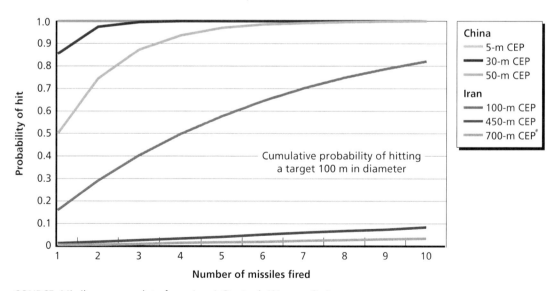

SOURCE: Missile accuracy data from *Jane's Strategic Weapon Systems*.
NOTE: Assumes perfect missile reliability, zero offset, and equal variance in *x* and *y* errors.
RAND *RR229-3.3*

accurate targeting data. Cruise missiles and armed unmanned aerial systems represent a growing threat to both fixed facilities and, increasingly, forces operating in the field.

This threat applies all of the services, particularly the Air Force, which operates far from fixed bases. For the cost of one ballistic missile, an opponent could purchase several long-range cruise missiles. The trade-off would be a greater number of less-survivable strike weapons. Because most cruise missiles are subsonic and fly at relatively low altitude, they are more vulnerable to defenses than are ballistic missiles. The same trade-off applies to UAVs in relation to ballistic missiles.

Maritime Anti-Access Capabilities

An important subset of long-range precision ballistic missiles is the new Chinese-developed ASBM. The emergence in the past few years of this new, potentially game-changing capability is a major concern to the U.S. Navy. The anti-ship cruise missile threat has been around for many years and is well understood by the Navy. Cruise missiles are normally subsonic, and most have ranges of less than 200 nautical miles, whether launched by aircraft, surface ship, submarine, or a shore-based coastal defense unit.

ASBMs have a much longer range than most anti-ship cruise missiles—perhaps over 2,000 km. They could be launched from mobile platforms toward an area where U.S. warships are operating and would probably use onboard sensors for the "end-game" attack on a warship.[3] China is currently working on this capability, with the centerpiece being the DF-21D missile. Should the entire system work as planned, it could force the U.S. Navy's surface elements to operate much farther from the Chinese coast than previous naval concepts of operation would have envisioned, or it could force the Navy to operate as currently planned but at much greater risk. Although ASBM technology is being pioneered by the Chinese military, it is possible that other nations could develop a similar capability in the future.

The ASBM system is vulnerable to complexities of the "kill chain," however. Unlike fixed land bases, such as airfields, ships move. Therefore, the Chinese would have to be able to (1) acquire the target ship, (2) identify it as a desirable target as opposed to a merchant ship or a decoy, (3) get the launch commands to a firing unit in a timely manner, (4) possibly provide in-flight target updates to the missile while it is en route, and (5) manage the end-game flight toward the target. For this process to succeed, the Chinese would have to overcome any active and passive defenses directed against the entire kill chain. Nevertheless, until and unless the Navy develops sufficient countermeasures to address this new threat, it could be forced to modify its concepts of operation in significant ways.

In littoral areas within a few hundred miles of a coast, there are a number of important A2 capabilities that an opponent could possess. These include high-quality non-nuclear submarines, fast missile-armed surface craft, and smart coastal and shallow-water mines. A number of nations have acquired modern diesel-electric or air-independent propulsion submarines. These very quiet designs can be armed with advanced torpedoes, mines, or submarine-launched cruise missiles. The quiet nature of their propulsion plants, coupled with sonar-absorbing or -reflecting hull coatings, makes these boats very difficult to detect. Despite

---

[3] Harry Kazianis, "Behind the China Missile Hype," *The Diplomat*, January 20, 2012.

the decades of effort devoted to anti-submarine warfare, the U.S. Navy could face a considerable A2 threat from this class of submarine.

Today's naval mines are far more sophisticated than the moored mines of World Wars I and II. Sea mines today can be deposited on the ocean floor with onboard sensors that can identify the most lucrative targets. Some mines will explode when triggered by their sensors, but others fire weapons, such as torpedoes. Because these systems are normally passive, detecting them before they fire can be a significant challenge for the Navy. It should be noted that some submarines are capable of laying mines in coastal waters. A related naval threat is unmanned undersea vehicles that can be used for reconnaissance and surveillance or to carry weapons.

A specialized form of naval mine are shallow-water weapons intended to threaten coastal traffic and vehicles or landing craft approaching a beach. Like other forms of naval mines, these weapons can be located and cleared, but this can be a very time-consuming process requiring days or even weeks. While it is unlikely that most opponents will be able to mine more than a fraction of their coastal areas, the mere presence of these threats can impose caution on naval operations, including the movement of ships being used to transport equipment and supplies for the Army.

The Proliferation of High-Quality Air Defenses

The air defense A2AD challenge can generally be broken into two types of threats. First, there is the medium-high-altitude threat that applies to aircraft operating above roughly 15,000 ft. In most cases, fairly large missiles are required to engage aircraft at those altitudes. The second type of air defense threat is low-altitude weapons—primarily shoulder-fired missiles (MANPADS) and anti-aircraft artillery. We address each threat type in turn.

Medium-High-Altitude Defenses

During the Yom Kippur War of 1973, the state-of-the-art Soviet-made air defense system that challenged the Israeli Air Force was the SA-6 mobile surface-to-air missile (SAM). The SA-6 had a maximum effective range of roughly 25 km. In the Gulf War in 1991, U.S. forces faced these and other longer-range but much older air defense systems. Today, the state of the art in air defense is the Russian-built S-400 (SA-21) SAM, which has already been sold to other countries. The S-400 has a maximum effective range against non-stealthy aircraft of approximately 400 km.[4] The S-400 is also highly resistant to electronic countermeasures, and, unlike the earlier SA-6, it can track and engage multiple targets. Importantly, today's modern SAMs (such as the S-400) are also very mobile. A well-trained SAM battery crew can disassemble the system and be ready to move within ten minutes. Coupled with a clever battle management system that manages radar emissions and integrated long-range S-300 (SA-20) and -400-type SAMs with short-range systems, these are formidable A2 capabilities. (Figure 3.4 shows a Russian-built S-300 system.)

The S-400 and similar systems are currently very expensive and require extensive training to implement successfully. This tends to limit the number of scenarios and the degree to which these systems will be confronted. That said, looking toward the future, these systems could represent an asymmetric approach, and one with significantly lower costs than developing and

---

[4] "S-400 SA-20 [sic] Triumf," Federation of American Scientists, last updated June 16, 2000.

Figure 3.4
Russian S-300 Surface-to-Air Missile System

SOURCE: Photo by Andrey Korchagin, used in accordance with Creative Commons
Attribution-NonCommercial-ShareAlike 2.0 licensing provisions.
RAND RR229-3.4

fielding an air force that is comparable to that of the United States. The U.S. Air Force and the U.S. Navy have yet to engage an opponent with this class of modern SAM. Until and unless the joint force can suppress this threat, U.S. air operations will be subject to major constraints.

Low-Altitude Air Defenses

As with long-range air defenses, such as the S-300/400, there has also been an increase in both the number and quality of lower-altitude air defense capabilities. Examples include MANPADS and larger, vehicle-mounted systems. Not every opponent can afford the latest generation of air defenses, but the cost of low-altitude defensive equipment is considerably lower than that of the larger S-300/400 or PATRIOT systems.

Today's modern air defenses are generally equipped with multiple types of sensors and fire-control systems, such as radar, electro-optical, and infrared capabilities, providing several options by which to acquire targets and engage hostile aircraft. If the target is a low-altitude aircraft, such as a transport plane loaded with paratroopers or a helicopter that is part of an air assault operation, these shorter-range air defenses can normally remain passive with their radars turned off until they are about to engage their targets. Therefore, while of a different nature than the S-300/400 threat, these weapons could also force airborne and air assault operations to be conducted at much greater distances from the objective area than has been the norm in the past.

The cumulative effect of an opponent's air defenses is, at a minimum, to impose caution on air operations. In the case of the low-altitude threat (MANPADS and anti-aircraft artillery), the ease with which these systems can be concealed means that they can suddenly show up in the rear area, such as at a key airfield, where they could pose an unexpected A2 challenge.

Low-altitude air defenses can also represent a significant AD threat to aircraft flying within the operational area. The Russian-made SA-15 (also known as the GK330 Tor; see Figure 3.5) represents this class of A2AD threat.

Long-Range Multiple-Rocket Launchers and Tactical Missiles

The post–World War II trend in artillery systems has been a pursuit of greater and greater range. In the 1970s, the main Soviet MRL was the truck-mounted 122-mm BM-21. That system had a maximum range of 20 km. Today, the norm in mobile multiple-rocket systems is a range of 80 km or more. For example, the Iranian 333-mm Fajr-5 has a range of roughly 75 km and has multiple warhead options, including the ability to carry high-explosive submunitions (see Figure 3.6).

This class poses a significant A2 challenge to Army forces. For example, today's airborne operations usually require the rapid seizure of a usable airfield where air-landed (as opposed to parachuted) follow-on forces and supplies can arrive to supplement the initial parachute units. However, if the airfield is within range of this class of enemy weapon, all operations there could be shut down. It should be noted that a number of long-range MRLs available today have much longer ranges than the Fajr-5. For example, the Chinese WS-2 heavy MRL has a projected range of 300–350 km and the ability to fire multiple warhead types, including submunitions. This long range provides the weapon with an operational-level reach.[5]

Figure 3.5
Russian Tor Air Defense System

SOURCE: Vitaly Kuzmin, used under the provisions of Creative Commons Attribution-ShareAlike 3.0 unported licensing provisions.
RAND RR229-3.5

---

[5] "The WS MRL Family," online discussion thread, *World Affairs Board*, last updated January 28, 2011.

Figure 3.6
Iranian Fajr-5 MRL

RAND *RR229-3.6*

Cyber Threats

In addition to traditional kinetic weapon threats, an opponent could pose a significant challenge to U.S. forces via cyber attacks. In the past two or three decades, the U.S. military has become very dependent on C4ISR systems, regardless whether it is engaging a state or nonstate opponent. The ability to acquire massive amounts of information, process it, and quickly disseminate it to large numbers of users over long distances is now a fundamental aspect of U.S. military operations.

Of the three types of opponents discussed in this report (irregular, hybrid, and state), the irregular opponent may have some degree of cyber capability, but it will probably be limited to occasional attempts to disrupt some U.S. operations. Some hybrid opponents may also be able to seriously disrupt the U.S. C4ISR system. Similarly, the ability of state opponents to degrade or disrupt the U.S. system will vary.

The important point is that, today and in the future, the U.S. military will be exposed to cyber threats that could be severe. It should be assumed that cyber attacks will be part of an opponent's A2AD operational approach.

Chemical and Biological Weapons

A potentially significant A2AD challenge is the threat of chemical and biological weapon use. Chemical weapons were used extensively in World War I and occasionally since then, including during the 1980s Iran-Iraq War. Biological weapons were stockpiled by some nations during the Cold War, but little if any use was made of them. While chemical and biological weapons do not have the instant, spectacular, and physically devastating effects of nuclear weapons, the threat or actual use of this class of weapon could serve to raise the stakes in a conflict and pose both A2 and AD challenges.

Another important characteristic of these weapons is that they are much easier to produce than nuclear weapons. Indeed, nonstate opponents (irregular or hybrid) could easily acquire or produce chemical or biological weapons. Delivery systems could include either conventional weapons (e.g., MRLs) or stealthy, clandestine means, such as packing them into commercial shipping containers, oil drums, or other cargo.

Nuclear Weapons: A Very Special Case

The most spectacular and potentially devastating military A2 capability that an opponent could possess is nuclear weapons. Fortunately, the number of countries with a nuclear arsenal is very small, and fewer still are potential opponents of the United States. That said, if the United States is on the verge of a crisis with a nuclear-armed state, the strategic and operational A2 threat might simply be perceived as too risky by senior U.S. decisionmakers.

There are several possible targeting strategies that a regional nuclear-armed state could consider. These "target sets" include the following:

1. *Warships at sea.* In some sense, this is the least risky strategy for a nuclear-armed regional power. A nuclear strike on ships at sea could be described as an attack on a purely military target, and there would be little if any collateral damage involving civilians or civilian infrastructure. However, for some regional nuclear powers, locating and targeting moving ships at sea might be too difficult. Much would depend on how long it would take to acquire the target, process that information and vet it with decisionmakers capable of authorizing a nuclear strike, send the necessary information to nuclear strike units (e.g., a nuclear-armed missile unit), and launch the attack. Ships move, and if this process requires hours, the target could be literally hundreds of miles from the location where it was detected.[6]

2. *Fixed military facilities ashore.* The best example is a military airfield, which has the advantage of being a non-moving target, unlike a warship. For some regional nuclear powers with a limited ability to conduct attacks against moving warships within an acceptable time frame, a fixed military facility could be a much more easily achievable target. However, a nuclear strike on an air base would, by definition, be on some nation's sovereign territory. Additionally, depending on whether there was a civilian community nearby, there could be massive amounts of collateral damage. As in the warship case, the justification would be that this was still a military target.

3. *Civilian infrastructure.* Examples include ports, oil refineries, or other civilian targets of value or, potentially, a direct strike on a population center. In the case of a port, an attack could have an immediate military effect as well, slowing or preventing the arrival of opposing military forces at that location. Such a move would almost certainly cause massive damage and possibly a huge number of civilian casualties. The threat of such a devastating strike might also cause a nation that is friendly to the United States to deny access by U.S. forces, particularly if the would-be host nation was convinced that the United States or other members of a coalition would be unable to protect it from such a strike. U.S. assurances of nuclear retaliation against a would-be perpetrator of

---

[6] John Gordon IV, Peter A. Wilson, John Birkler, Steven Boraz, and Gordon T. Lee, *Leveraging America's Aircraft Carrier Capabilities: Exploring New Combat and Noncombat Roles and Missions for the U.S. Carrier Fleet*, Santa Monica, Calif.: RAND Corporation, MG-448-NAVY, 2006, pp. 48–50.

the nuclear attack might not be good enough for the threatened nation, given the risks involved.

4. *U.S. military forces operating inside the adversary country.* In this case, the nuclear-armed nation waits for U.S. or other coalition forces to enter its territory and then conducts a nuclear strike against them. Public justification for such a move would probably be framed as "self-defense" against an invader. The threatened nation that makes such a strike on its own territory might be willing to accept a degree of collateral damage to inflict massive casualties on the invading forces. It is possible that a nuclear first use under these conditions would be accompanied by a threat of further nuclear use (as in the third case discussed) should the United States retaliate with nuclear weapons.

5. *A high-altitude nuclear burst to disrupt U.S. C4ISR systems.* A nuclear burst several hundred kilometers above the earth could severely disrupt or permanently disable many electronic systems in a region. This action could be taken without inflicting any direct casualties and with little or no damage to physical infrastructure. In that sense, this might be the most appealing option to an opponent. Additionally, an initial high-altitude burst could serve as a stern warning that further nuclear use might follow, as in the third and fourth cases.

These five illustrative situations do not include all the possible targeting strategies of a nuclear-armed opponent. Nuclear weapons represent the high end of military A2 capability that will require the United States to operate differently if these systems are in the hands of an adversary.[7]

---

[7] David Ochmanek and Lowell H. Schwartz, *The Challenge of Nuclear-Armed Regional Adversaries*, Santa Monica, Calif.: RAND Corporation, MG-671-AF, 2008, pp. 15–40.

CHAPTER FOUR

A Joint Approach to Countering Anti-Access/Area Denial Challenges

As discussed earlier in this report, U.S. forces could face a wide range of A2AD threats. A review of U.S. operations conducted since World War II, combined with the development and examination of possible future scenarios, indicated that, in most situations, the United States would employ a joint approach to overcoming diplomatic, geographic, and military A2AD challenges. The capabilities of the different services can be brought to bear in ways that are tailored for each specific operational environment.

In the area of *diplomatic* challenges, actions taken by the U.S. military can offer considerable assistance. During periods of peace, when there is normal engagement (often referred to as "Phase 0" activities), U.S. military activities can improve the likelihood of access in the event of a crisis. Indeed, the peacetime presence of the U.S. military might, in itself, negate or dramatically reduce the possibility of a military crisis in a particular region. Indeed, pre-crisis activities and the presence of the U.S. military can be beneficial in overcoming A2 challenges, particularly in the diplomatic realm:

- The presence of U.S. military forces demonstrates American interest and intent to protect friends and allies.
- There is an opportunity to train and exercise with the armed forces of friends and allies in a region to learn one another's procedures. Such engagement will help minimize coalition operational challenges and maximize interoperability should a crisis occur.
- Key pieces of equipment, such as munitions and unit supplies, can be prepositioned ashore or afloat, thus reducing the amount of tonnage that has to be deployed in the event of a crisis.
- Senior U.S. military personnel and planners can become familiar with regional issues, such as which infrastructure is available to facilitate the arrival of U.S. and coalition forces in the event of a crisis.

The size and role of U.S. forces in a particular country or region will vary, often as a result of domestic political concerns in individual nations. Those realities can also influence which service is best suited for peacetime, or Phase 0, engagement activities. Some countries will allow the United States to station an appropriately sized military force inside their borders during peacetime. In some cases, this could include sizable U.S. ground forces, as in the Republic of Korea; in others, it may be air and naval forces, as in Japan. In many host countries, internal politics will allow only periodic exercises with U.S. forces. The Cobra Gold exercises with the armed forces of Thailand are a good example.[1] There is no U.S. military force permanently stationed in West Africa, but the U.S. Navy sends ships for months at a time to the Africa Partnership Station to train police and military forces from various West African nations and to improve interoperability. These ships can operate offshore or anchor at regional ports.

The U.S. Army provides special value in several respects during Phase 0 efforts to improve access for U.S. forces in a country or region. In most countries, the army dominates the military force. There are a number of reasons for this, not the least of which is that air forces and navies represent large capital investments that many nations cannot afford. Unfortunately, some countries that are friendly to the United States are confronted with internal security problems, including insurgencies. In those cases, the countries' armies (and police forces) tend to be the most important security contingent.

For these reasons, the U.S. Army can be particularly useful in engaging the army-dominated militaries of many nations. As mentioned earlier, there are many situations in which the permanent assignment of U.S. Army forces inside a country during peacetime is not possible, usually due to political concerns in the United States or the host nation. In those situations, small training teams periodically deployed to the host country, allowing key leaders from the host-nation military to attend U.S. Army schools, and periodic combined exercises with that country's army can be of mutual benefit—and will contribute to gaining access in the event of a major crisis.

In terms of overcoming *geographic* access challenges, the Army is, to a large extent, dependent on the capabilities of the Air Force and Navy—including the future modernization plans of those services. Unless a deployment can be accomplished with leased commercial assets, the Army generally requires Air Force and Navy capabilities. The Army has organic aircraft (mostly helicopters) and some coastal shipping assets, but they are limited in both capacity and range. Thus, the Army is heavily reliant on the Navy and Air Force, which have the majority of the capability required to move the Army. As the Army examines ways of reducing the effect of geographic challenges to access, the continued prepositioning of Army unit equipment sets and supplies, both ashore and afloat, should be examined. During the Cold War, this was a very useful technique to expedite the Army's deployability. Additionally, it may be possible to improve access to commercial air- and sealift to help deploy the Army, a move that would relieve some of the burden on Air Force and Navy aircraft and ships.

In the case of the Air Force, the service's buy of intercontinental-range transport aircraft is just about finished. There are a total of 213 C-17 and 79 C-5 transport aircraft in the airlift fleet.[2] The Air Force provides the capability to rapidly transport small Army formations (roughly one brigade at a time). Importantly, the Air Force can deliver and sustain Army elements far inland, a great distance from ports. For intratheater air mobility, the Air Force's C-130 force is of considerable help to the Army, although it should be noted that the C-130 is not considered a long-range intertheater transport aircraft. As of this writing, there were 428 C-130s of various models in the Air Force.[3]

---

[1]   Donna Miles, "Cobra Gold 2012 to Promote Partnership, Interoperability," American Forces Press Service, January 13, 2012.

[2]   U.S. Air Force, "C-17 Globemaster III," fact sheet, December 29, 2011a; U.S. Air Force, "C-5 A/B/C Galaxy and C-5M Super Galaxy," fact sheet, August 16, 2012.

[3]   U.S. Air Force, "C-130 Hercules," fact sheet, December 29, 2011b.

The Navy (including the option to lease commercial shipping assets) has the capability to transport a huge quantity of tonnage. Since the Vietnam War, the norm has been for Army personnel to fly to an operational area in Air Force transport aircraft or commercial airliners, but the vast majority of the Army's equipment and supplies travels by sea. During the Gulf War, more than 95 percent of all tonnage deployed to the theater went by ship. A single large medium-speed roll-on/roll-off ship (LMSR) can carry roughly the same amount of tonnage as all the C-17s and C-5s in the Air Force inventory.[4] What the Navy and leased commercial ships need is usable and protected ports. Some modern ships are so deep that they can use only a limited number of ports. In other cases, the facilities at ports are so poor that the offloading of cargo can be significantly delayed.

Overcoming Military Anti-Access Challenges

Military A2 challenges are an area of particular interest to the U.S. armed forces. In that regard, there are important relationships and interdependencies among the services. These interdependencies highlight how A2 threats to one service can affect the others. In this section, we discuss how an evolving military A2 threat might affect each service's future operations, with emphasis on threats posed by better-armed opponents and characteristic of the hybrid and state adversaries—the types of capabilities that were highlighted in Chapter Three.

Since the Goldwater-Nichols legislation of the 1980s, the U.S. military has taken an increasingly joint approach to operations.[5] A joint approach is particularly important to overcoming A2AD threats. The following sections describe some of the relationships between the Army and the other services in terms of managing or eliminating A2AD challenges.

Anti-Access Challenges to the Air Force

The development of precision ballistic and cruise missiles represents a major threat to U.S. Air Force operations. With the introduction of ballistic and cruise missiles with ranges of well over 1,000 miles and CEPs measured in tens of meters, the nature of the threat to land-based air operations has changed profoundly. This class of weapon has a high probability of damaging runways or destroying aircraft on the ground unless the air base is well defended, such as with hardened aircraft shelters and runway repair equipment.

Against a country like China, which has a large and growing inventory of precision ballistic and cruise missiles, the Air Force will be forced to (1) operate from greater distances, knowing that most opponents will have fewer very long-range missiles; (2) disperse its operations to a larger number of bases, forcing an opponent to expend its missile arsenal against more targets, each with fewer aircraft than a large base; or (3) to the extent possible, increase the defenses of each of its bases. In that regard, the defenses against cruise missiles are very different from those needed to counter the ballistic missile threat. The former is a much easier technical challenge, and defense strategies tend to be far cheaper than those for ballistic missile threats.

If the Air Force is compelled to operate farther away to minimize the missile threat to its bases, the arrival location of Army personnel and units will also be farther from the immediate

---

[4] Lynn E. Davis and Jeremy Shapiro, eds., *The U.S. Army and the New National Security Strategy*, Santa Monica, Calif.: RAND Corporation, MR-1657-A, 2003, pp. 198–200.

[5] See Public Law 99-433, Goldwater-Nichols Department of Defense Reorganization Act of 1986, October 1, 1986.

area of operations. This could complicate and slow the initiation of important Army capabilities in the region. Additionally, the farther away from the operational area that the Air Force must operate, the lower the sortie rate of its aircraft. For Army forces engaged in a theater, the amount and timeliness of Air Force support could be adversely affected by the need to base outside the range of all or most enemy missiles.

How the Army Can Help Enable Air Force Operations

Given the A2AD threat to Air Force bases, the Army could also have a role in providing protection. This could come in three forms. First, there could be a need for active missile defenses at Air Force bases, particularly if those bases are located inland, where the Navy would not be able to provide protection with its Aegis-armed cruisers and destroyers. Active defense is less of a challenge against cruise missiles than against ballistic missiles. It is possible to defend against cruise missiles with an appropriate mix of early warning/surveillance radars, such as Sentinel, and defensive missiles, point-defense guns, and, perhaps, speed-of-light weapons, such as lasers or directed-energy systems. We address the Army's possible role in greater detail later in this chapter.

Second, if there is a ground threat to forward Air Force bases in the operational area, the Army might have to provide ground maneuver units in defensive roles, particularly if the host nation is unable to provide adequate protection. Such threats could consist of commando-type units or conventional units, such as air assault or airborne forces. The threat could include indirect fire from artillery or rockets.[6]

Third, the Army could provide engineering and logistics support. If the Air Force is forced to operate from dispersed locations, it might need the Army's assistance to bring fuel, munitions, and other supplies to those bases, especially if air bases are located inland, away from ports. The Air Force's possible dependence on the Army for logistics assistance would be particularly strong if it were operating from bases where few or no supplies and munitions were already positioned. Army engineering support could be needed if bases required improvements or repair due to enemy missile bombardment, especially given the time constraints (and cycle times for regeneration) that would likely be involved.

The Army's Role in Suppressing Air Defenses

The proliferation of high-quality air defenses is another significant A2AD challenge for the Air Force (and naval aviation). Chapter Three discussed the types of systems that pose a major challenge to air operations. The long-range systems of the S-300/400 type will inspire caution on the part of commanders charged with decisionmaking for all types of aircraft; this threat could force large subsonic, non-stealthy cargo and surveillance platforms to operate far from the threat or require a major change in tactics (e.g., non-stealthy transport planes would have to conduct more very-low-altitude operations to avoid radar detection).

The Army can provide capabilities to destroy or suppress enemy air defenses in support of Air Force and Navy aviation. There is historical precedent for this kind of ground force support to air forces. During the 1973 Yom Kippur War, once Israeli ground forces reached the west bank of the Suez Canal, they quickly deployed long-range 175-mm self-propelled guns against their bridgehead. Those guns were used to engage Egyptian SAM batteries, thus reducing the

---

6 Alan J. Vick, *Snakes in the Eagle's Nest: A History of Ground Attacks on Air Bases*, Santa Monica, Calif.: RAND Corporation, MR-553-AF, 1995.

overall air defense threat to Israeli aircraft. Once Israeli ground forces began to rapidly expand their bridgehead, they overran numerous fixed Egyptian SAM sites and forced mobile SAM units to flee, lest they also be overrun by Israeli armored units. In the 1991 Gulf War, the Army Tactical Missile System (ATACMS), equipped with anti-personnel, anti-materiel bomblets, was used to attack and suppress enemy air defense sites.

Depending on the specifics of a future crisis, the Air Force and Navy could benefit from such Army capabilities as long-range (up to roughly 300 km) ATACMS Block II artillery fires (see Figure 4.1) to engage enemy air defenses, as well as the ability of Army ground forces to overrun or drive off hostile SAMs.[7] Army unmanned aerial systems might be useful for target location and surveillance of hostile air defense systems to alert and guide deep-strike Army or joint systems toward the target. In light of the longer-range threat systems that are proliferating, the Army may need even longer-range deep-strike systems. Regardless, these are examples of the joint interdependencies that could come into play when confronting enemy A2 threats.

Anti-Access Challenges to U.S. Naval Forces

As mentioned in Chapter Three, ASBMs and cruise missiles represent a major threat to U.S. naval operations. The ASBM threat, in particular, is different from the decades-old anti-ship cruise missile threat that the Navy has faced since the 1960s. The much longer range of the ASBM challenge could require U.S. naval forces to operate much farther from an enemy coast than in the past. In littoral areas, mines, fast-attack craft, and fast suicide craft employing swarming tactics, along with submarine threats, are all major concerns to U.S. naval forces. Amphibious operations by the Marine Corps could be seriously affected by these threats. If the

Figure 4.1
MGM-140 ATACMS

SOURCE: U.S. Army photo.
RAND *RR229-4.1*

---

[7] ATACMS Block I was used against Iraqi air defense capabilities in the Gulf War in 1991.

Navy cannot develop adequate countermeasures to various littoral threats, its ability to operate ships close to the coast could be seriously compromised.[8]

From the Army's perspective, these A2 capabilities that threaten naval operations could have several effects. First, should a combination of hostile submarines, ASBMs and long-range anti-ship cruise missiles, and enemy aircraft force the U.S. Navy to operate farther from the objective area, by definition, it will be more difficult (if not impossible) to deploy Army forces by sea. Related to the first concern, even if Army forces can be deployed into the operational area, perhaps in relatively small numbers, the Navy might not be able to remain sufficiently close to the Army's operational area to provide timely and effective fire and air support or to meet the Army's logistical needs.

The littoral threats could constrain the arrival of Army (and Marine Corps) forces in the operational area, whether through sea-based forced entry or an administrative move ashore from shipping. Shallow-water mines could impose significant threats to Marine Corps amphibious operations. Even after a forced entry operation has theoretically cleared a portion of enemy coastline, it is conceivable that the area offshore could be "reseeded" with high-tech naval mines delivered by enemy submarines or unmanned undersea vehicles. High-speed missile-armed vessels that can hide in cluttered terrain along a coast or mingle with civilian shipping can also pose a serious threat to U.S. ships operating in littoral regions.

Army Support to Naval Forces

As was the case with the Air Force, there are situations in which the Navy could benefit from the Army to overcome or reduce enemy A2 capabilities. When fighting a major opponent that has a significant offensive missile capability, it is likely that the Navy's Aegis-armed cruisers and destroyers will be in high demand to protect the fleet, help shield the other services' elements operating in the littoral region, and help defend cities and other economic assets of U.S. allies in a region.

Today's *Arleigh Burke*–class destroyers are armed with 90 vertical-launch system tubes. Some of those tubes are devoted to offensive strike weapons, such as the Tomahawk Land Attack Missile, while others are set aside for defensive missiles, such as the Standard Missile 2, 3, and 6. When a destroyer (or Aegis cruiser) needs to reload its missiles, it must retire to a port and spend a day or more with access to cranes to lift new missiles into the launch tubes; reloading cannot be done at sea. Because the ship is stationary for the entire reloading cycle, it is potentially vulnerable. That could cause the Navy to establish rearm locations well outside the range of enemy missiles. While such a move would help ensure the safety of the ship during the vulnerable reloading period, it also means that the ship might have to steam many hundreds of miles outside the operational area to find a safe reloading site. If, on the other hand, the Army could deploy missile defenses to a closer location—perhaps not completely beyond the entire enemy missile "threat ring" but outside the range of the most threatening missiles— the total time that the ship would be unavailable for action could be cut by several days.

In terms of the threat posed by enemy missile-armed littoral surface craft and small fast-attack suicide boats, the Army may be able to assist the Navy by providing unmanned aerial systems or attack helicopters to help locate and eliminate enemy craft operating in the coastal

---

[8] It should be noted that although the majority of amphibious operations since World War II have been conducted by the Marine Corps, the Army has and will continue to play a role in sea-based operations, both to supplement Marine Corps forces and, on occasion, to lead operations (as was the case in Haiti in the 1990s).

regions. Depending on how large an area must be patrolled, and the amount of clutter in a particular coastal region, a considerable number of aerial patrol assets could be required. There is precedent for the use of Army attack helicopters in such a role. U.S. Army Apaches have been used to patrol waters of the Arabian Gulf, and as recently as 2011, British Army Apaches destroyed Libyan coastal patrol boats using Hellfire missiles.[9]

Finally, there could be situations in which Army watercraft could be a useful supplement to Navy and Marine Corps shallow-water vessels. For example, if the main ports in a region are unusable, the ability to use minor ports or over-the-shore logistics could be critical.

Options to Reduce Anti-Access Threats to the Army

There are a variety of A2 threats that could be especially problematic for the Army. Some of them, such as the development of increasingly long-range air defenses, are also a challenge for the Air Force and Navy. Others are more specific to the Army. In this section, we provide an overview some of the most serious threats against which the Army will have to develop counters, either on its own or in concert with the other services.

Mitigating the Air Defense Threat

Chapter Three highlighted two major classes of air defense threats: medium-high-altitude and low-altitude. In general terms, the Air Force and naval aviation pay more attention to eliminating or suppressing the former. The development of modern air-delivered precision munitions and high-quality intelligence, surveillance, and reconnaissance (ISR) systems have greatly reduced the need for Air Force and Navy strike aircraft to operate at low altitude; today, they can achieve great accuracy from above 15,000 ft. Therefore, the Air Force and Navy tend to focus on the medium-high-altitude threats.

From the Army's perspective, both classes of air defenses have important implications. Unless eliminated or suppressed, air defense may render the Army's premier forcible-entry capabilities—airborne forces, such as the 82nd Airborne Division—unable to exercise their core mission. It might be too risky for fighters to operate within the range of these AD assets, let alone the slower, non-stealthy transport planes (e.g., C-17 or C-130) that are required by Army airborne forces.

If an opponent has long-range SAMs, such as the S-300/400, the Army will probably have to conduct airborne operations at much greater distances from the objective than in the past. While the Air Force and Navy can be expected to devote considerable effort to locating and suppressing these weapons, there could be a residual threat of a few enemy batteries hiding with their radars turned off until an appropriate target—such as a large transport aircraft—appears. If there is even suspicion that such a threat still exists, air planners will probably insist that the drop zones for Army airborne forces be located as far from the threat as possible. This was the situation in Kosovo in 1999 during Operation Allied Force, when NATO air commanders were never sure whether the SA-6s operating inside Kosovo had been eliminated because of the Serbs' efforts to hide their air defense assets.[10] Given the long range (well over

---

[9] Philip Ewing, "U.K. Apaches Destroy Two Libyan Vessels, AA Gun," *DoD Buzz*, June 14, 2011.

[10] Bruce R. Nardulli, Walter L. Perry, Bruce R. Pirnie, John Gordon IV, and John G. McGinn, *Disjointed War: Military Operations in Kosovo, 1999*, Santa Monica, Calif.: RAND Corporation, MR-1406-A, 2002, pp. 28–30.

100 km) of these air defense systems, this could impose significant operational limitations on Army airborne operations, particularly because today's airborne infantry units are generally foot-mobile, as they were in World War II.

What can the Army do to reduce the severity of these threats? In most cases, the actual suppression of enemy air defenses will likely be the responsibility of either the Air Force or naval aviation. As mentioned earlier, in some scenarios, the Army can contribute to the suppression effort with its systems, such as the long-range ATACMS. Additionally, it could modernize some or all of its airborne units to give them an air-droppable light mechanized/motorized capability. This would allow some airborne units to drop in much farther from the worst of the enemy's air defenses and then move toward objective areas with motorized or light mechanized units. Of course, increasing the number of vehicles in an airborne unit will require that more transport aircraft be devoted to the operation. Figure 4.2 shows the Marine Corps' LAV-25 light armored vehicle, a good example of the type of air-droppable system that could be useful to Army airborne units.

It is useful to point out that an air-droppable light-armor concept bears some resemblance to the "air mechanization" and "mounted vertical maneuver" concepts that were closely associated with the Army's Future Combat Systems (FCS) program, which was canceled in 2009. Conceptually, there is some similarity in the sense that early-deploying light armor arriving by air was a key feature of FCS. In this case, we envision that the kind of enhancement mentioned here would be implemented in only some of the Army's airborne units—perhaps only a single brigade. (There were six airborne infantry brigades as of as of this writing.) In contrast, the FCS program had much more ambitious goals, targeting as much as one-third of the active Army.

Figure 4.2
U.S. Marine Corps LAV-25 Light Armored Vehicle

SOURCE: U.S. Marine Corps photo by LCpl Kamran Sadaghiani.
RAND *RR229-4.2*

The key features of this proposed enhancement to today's airborne forces would be as follows: (1) it would provide a degree of mounted combat capability to airborne units, allowing them to land farther from their objectives than today's generally foot-mobile airborne units, thus avoid some of the most severe A2 threats; (2) it would utilize the current U.S. Air Force transport aircraft fleet (in contrast, FCS envisioned acquiring large numbers of vertical-takeoff-and-landing aircraft for the Army); and (3) it would involve equipping only a portion of the airborne force with this capability because of the limited number of available Air Force transport planes.

Another technique the Army could use to help reduce the effect of improved enemy air defenses is to make greater use of the Joint Precision Air Drop System (JPADS; see Figure 4.3). JPADS would allow Air Force cargo aircraft to remain well above low-altitude air defenses, such as anti-aircraft guns and MANPADS, and still deliver equipment and supplies (though not personnel) from high altitude with great accuracy. JPADS is a GPS-based system that can deliver loads of up to five tons to within 30–50 m of the designated landing point. (The follow-on JPADS-M system will be able to deliver 15–30 tons.) Coupled with the use of light armored vehicles in some airborne units, JPADS could resupply those units once they start advancing toward their objectives, as it did in Afghanistan.

The Long-Range Artillery Threat

Today, the Army is limited in its ability to respond to the long-range artillery threat. Its own Multiple Launch Rocket System and High Mobility Artillery Rocket System are limited to a range of 84 km. While the ATACMS has a much greater range, quickly transporting a significant number of these bulky, heavy missiles to a just-seized airfield could be difficult. The

Figure 4.3
JPADS Delivering Cargo

SOURCE: U.S. Air Force photo.
RAND *RR229-4.3*

155-mm howitzers organic to airborne units are limited to roughly 30-km range. This apparent capability gap is an area that the Army should address, both internally and in concert with the other services.

Cruise Missiles and Unmanned Aerial Vehicles

The A2 threat posed by cruise missiles and UAVs is an area in which technology could serve the Army well. There are a number of options that the Army could consider to increase its defenses against this type of threat. In general terms, weapons are either air defense guns and missiles or directed-energy weapons, such as lasers. During the past decade's operations in Iraq, a version of the U.S. Navy's Close-In Weapons System (see Figure 4.4), a 20-mm rapid-fire gun, was used to defend bases from rocket attacks. That interim solution represents the kind of gun system that the Army could further develop. Another, longer-term option would be truck- or trailer-mounted mobile laser or directed-energy systems that could engage cruise missiles or UAVs. With all of these defenses, streamlined C2 systems are needed to quickly pass intelligence to units that are at risk, perform combat identification, assign sensors and weapons, manage the airspace, and establish rules of engagement. New C2 architectures, such as the Joint Air-Ground Integration Cell, should shift more of these functions to the ground sectors.

Not only could this capability be valuable to protect Army forces, but it could also be very useful for the defense of joint assets, such U.S. Air Force bases and the important economic infrastructure of U.S. allies in a region. As in the example of threats to the airfields that are seized as part of an airborne operation, if these systems were sufficiently mobile, they could be quickly deployed to provide defense against rocket or UAV attacks.

Figure 4.4
U.S. Navy Close-In Weapons System Aboard the USS *Monterey*

SOURCE: U.S. Navy photo by MC3 William Weinert.
RAND *RR229-4.4*

Combining Ground Maneuver with Other Joint Capabilities

Many of the A2 threats discussed here have the effect of inhibiting the introduction of ground forces into an operational area. Threats to Air Force aircraft and Navy ships can delay or prevent the arrival of Army and Marine Corps ground forces. Threats to air bases can degrade the amount and timeliness of air support to ground forces. That said, in some scenarios, the introduction of ground forces could actually help overcome the A2 threat to joint forces.

An example is the case of Iran attempting to close the Strait of Hormuz with a combination of mines, coastal defense missiles and guns, high-speed attack boats, and aircraft. Should the Iranians actually try to do this, it is likely that the United States and its allies would attempt to eliminate those threats with precision-strike munitions, delivered primarily by aircraft. Should the use of fires alone be insufficient to eliminate the threat to shipping vessels passing through the strait, introducing an appropriately sized ground force could force the Iranians to move, and therefore expose, their anti-ship missile launchers and air defense units operating in the vicinity. The fact that their units would be moving to avoid being destroyed by ground forces would make them more easily detectable by other elements of the joint force, and probably much easier to target and destroy. Additionally, the presence of U.S. ground forces in the area would minimize the likelihood that Iranian forces could deploy replacement units to reestablish a threat to shipping.

Increased Use of Unmanned Systems

The Army has already made a major commitment in the area of unmanned air and ground systems. In the past decade, the needs of troops in combat in Iraq and Afghanistan have accelerated the move toward the increased use of these systems.

Unmanned systems could benefit the Army and the other services as they confront many of the A2AD threats discussed in this report. For example, armed unmanned aerial systems might be of considerable value against both medium-high- and low-altitude enemy air defenses. The maritime surveillance mission, in which considerable littoral areas require constant surveillance, could be an ideal mission for long-endurance unmanned systems. Unmanned ground systems for both ISR and attack could be particularly useful in high-threat areas. Unmanned ground systems would also have the advantage of providing persistent coverage of an area, perhaps for days or even weeks.

As the Army considers possible solutions to address the evolving A2AD threat, greater use of unmanned systems could be one of the most promising areas for technology solutions.

Just as the Army can be of assistance to the Air Force and Navy in overcoming A2 threats to their operations, the other services can assist the Army (and Marine Corps) in eliminating or reducing the severity of AD threats to ground force operations.

In many operations, a large proportion—or even a majority—of airborne ISR systems will belong to the Air Force and Navy, particularly long-range, high-altitude platforms. As operations in Iraq and Afghanistan have demonstrated, these long-endurance unmanned systems can greatly increase the situational awareness of U.S. ground forces. In future operations, this will probably still be the case, although the degree of the air defense threat will have important effects on U.S. airborne ISR systems.

CHAPTER FIVE

Conclusions

The terrorist attacks of September 11, 2001, and the global war on terrorism took the services and their weapon modernization initiatives in a new direction. From 2002 to 2012, the Army shifted much of its portfolio of capabilities to better address the lower end of the threat spectrum. Meanwhile, global threats have continued to evolve. As the U.S. military looks toward the future, it must address the A2AD challenges that it currently faces and those that are likely to arise in the coming years. Importantly, the military writ large, and the Army in particular, will have to cope with a wide range of A2AD challenges, from low- to high-end capabilities.

The Army's experience over the past decade should leave it relatively well prepared to address certain potential A2AD threats, particularly those that involve insurgency-like operations and tactics. For example, the Army now has considerable experience with the IED threat that will probably be a common AD technique used by future opponents. However, the advancement of threat capabilities at the middle and higher end of the threat spectrum means that many potential adversaries are a much greater challenge now than they were a decade ago, particularly in terms of denying U.S. forces access to an operational area. For example, ASBMs were not recognized as a threat when the U.S. military first became heavily engaged in counterterrorism and counterinsurgency operations early in the last decade.

Ironically, the A2 challenge now appears to be exacerbated by the cancellation or postponement of key capabilities that the Army was in the process of developing as the Cold War ended. Many of these advanced weapon and modernization programs, designed to defeat a much larger Soviet Union over the northern plains of Germany, were considered obsolete. And while the massive adversary force no longer exists, many of its key capabilities have proliferated, either in exact form or in copy. Examples include high-end, high-altitude air defenses; MANPADS; high-volume, long-range artillery; advanced tactical weapons (e.g., shaped charges, explosively formed projectiles); and nuclear weapons. As technology makes its mark, guidance for missiles (both ballistic and cruise), along with variety of unmanned systems and other advancements, can be added to the mix.

In the A2AD environment, perhaps the total force size matters less than *force capability*, at least initially. Unlike the Cold War era, in which large numbers of forces were already prepositioned in theater, in the modern A2AD scenario, U.S. forces may be outnumbered by a much greater margin as they attempt to gain and maintain access. When force ratio and capability disparity are combined with current and continued threat advancement, the challenge to the joint forces can be substantial. In the scenarios we examined for this study, there was a high premium on advanced force protection capabilities, particularly for protecting U.S. forces from an opponent's A2 capabilities. These capabilities include both defensive systems, such as the C-RAM (counter–rocket, artillery, and mortar) system, and offensive options, such as

33

long-range fires. Regardless of the means of entry into a theater, force protection is an essential U.S. capability and will serve as an enabler for force buildup. Thus, establishing a sufficiently high force ratio and rate of inflow early on—and providing a means to protect that force—can facilitate access to a theater and help ensure that the area of operations is a location of U.S. combatant commanders' choosing.

There are important interdependencies and synergies between the services in terms of their ability to overcome A2AD challenges. For example, some threats to the Army's ability to deploy to an operational area are addressed primarily by the other services. In most cases, A2AD threats require joint solutions.

The Army has capabilities that can be of great value to the Air Force and Navy as they develop joint solutions to overcoming evolving A2 challenges. Specifically, the Army could provide considerable assistance to the Air Force and Navy in suppressing or destroying air defenses that challenge joint air operations. It could also facilitate dispersed operations by the Air Force to assist the Navy in defeating littoral threats and establishing protected regional enclaves for naval operations. Army ground-based air defense systems (including today's PATRIOT and next-generation systems to protect against UAVs, cruise missiles, and other threats) could be critically important for the protection of critical joint assets, such as ports and airfields.

As the services work together to develop operational concepts and systems to mitigate A2AD challenges, the Army will play a significant role in this effort.

Bibliography

"China Building an Army of Unmanned Military Drones 'to Rival the U.S.,'" *Daily Mail*, July 5, 2011. As of January 29, 2013:
http://www.dailymail.co.uk/news/article-2011533/China-building-army-unmanned-military-drones-rival-U-S.html

"Chinese Develop Special 'Kill Weapon' to Destroy U.S. Aircraft Carriers," U.S. Naval Institute, March 31, 2009. As of January 29, 2013:
http://www.usni.org/news-and-features/chinese-kill-weapon

Davis, Lynn E., and Jeremy Shapiro, eds., *The U.S. Army and the New National Security Strategy*, Santa Monica, Calif.: RAND Corporation, MR-1657-A, 2003. As of January 29, 2013:
http://www.rand.org/pubs/monograph_reports/MR1657.html

"DF-15/-15A/-15B (CSS-6/M-9)," *Missile Threat*, last updated January 5, 2013. As of January 29, 2013:
http://missilethreat.com/missiles/df-15-15a-15b-css-6m-9/

Ewing, Philip, "U.K. Apaches Destroy 2 Libyan Vessels, AA Gun," *DoD Buzz*, June 14, 2011. As of January 29, 2013:
http://www.dodbuzz.com/2011/06/14/u-k-apaches-destroy-2-libyan-vessels-aa-gun

Gordon, John IV, Peter A. Wilson, John Birkler, Steven Boraz, and Gordon T. Lee, *Leveraging America's Aircraft Carrier Capabilities: Exploring New Combat and Noncombat Roles and Missions for the U.S. Carrier Fleet*, Santa Monica, Calif.: RAND Corporation, MG-448-NAVY, 2006. As of January 29, 2013:
http://www.rand.org/pubs/monographs/MG448.html

Hoffman, Frank G., "Hybrid Warfare and Challenges," *Joint Force Quarterly*, No. 52, 1st Quarter, 2009, pp. 34–39.

International Institute for Strategic Studies, *The Military Balance 1988–1989*, London, 1988.

———, *The Military Balance 2010*, London, February 2010.

Johnson, David E., *Hard Fighting: Israel in Lebanon and Gaza*, Santa Monica, Calif.: RAND Corporation, MG-1085-A/AF, 2011. As of January 29, 2013:
http://www.rand.org/pubs/monographs/MG1085.html

Kazianis, Harry, "Behind the China Missile Hype," *The Diplomat*, January 20, 2012. As of January 29, 2013:
http://thediplomat.com/2012/01/20/behind-the-china-missile-hype

Kopp, Carlo, "Almaz S-300—China's 'Offensive' Air Defense," International Assessment and Strategy Center, February 25, 2006. As of January 29, 2013:
http://www.strategycenter.net/research/pubID.93/pub_detail.asp

Krepinevich, Andrew F., Jr., *The Army and Vietnam*, Baltimore, Md.: Johns Hopkins University Press, 1988.

Miles, Donna, "Cobra Gold 2012 to Promote Partnership, Interoperability," American Forces Press Service, January 13, 2012. As of January 29, 2013:
http://www.defense.gov/news/newsarticle.aspx?id=66803

Nardulli, Bruce R., Walter L. Perry, Bruce R. Pirnie, John Gordon IV, and John G. McGinn, *Disjointed War: Military Operations in Kosovo, 1999*, Santa Monica, Calif.: RAND Corporation, MR-1406-A, 2002. As of January 29, 2013:
http://www.rand.org/pubs/monograph_reports/MR1406.html

National Air and Space Intelligence Center, *Ballistic and Cruise Missile Threat*, Wright-Patterson AFB, Ohio, NASIC-1031-0985-09, 2009.

Ochmanek, David, and Lowell H. Schwartz, *The Challenge of Nuclear-Armed Regional Adversaries*, Santa Monica, Calif.: RAND Corporation, MG-671-AF, 2008. As of January 29, 2013:
http://www.rand.org/pubs/monographs/MG671.html

O'Connor, Sean, *PLA Ballistic Missiles*, Air Power Australia, Technical Report APA-TR-2010-0802, August 2010. As of January 29, 2013:
http://www.ausairpower.net/APA-PLA-Ballistic-Missiles.html

Office of the Secretary of Defense, *Unclassified Report on Military Power of Iran*, Washington, D.C., April 2010.

———, *Annual Report to Congress: Military and Security Developments Involving the People's Republic of China*, Washington, D.C., 2011. As of January 29, 2013:
http://www.defense.gov/pubs/pdfs/2011_CMPR_Final.pdf

Public Law 99-433, Goldwater-Nichols Department of Defense Reorganization Act of 1986, October 1, 1986.

"R-11/SS-1B SCUD-A, R-300 9K72 Elbrus/SS-1C SCUD-B," Federation of American Scientists, last updated September 9, 2000. As of January 29, 2013:
http://www.fas.org/nuke/guide/russia/theater/r-11.htm

"S-400 SA-20 [sic] Triumf," Federation of American Scientists, last updated June 16, 2000. As of January 29, 2013:
http://www.fas.org/nuke/guide/russia/airdef/s-400.htm

U.S. Air Force, "C-17 Globemaster III," fact sheet, December 29, 2011a. As of January 29, 2013:
http://www.af.mil/information/factsheets/factsheet.asp?id=86

———, "C-130 Hercules," fact sheet, December 29, 2011b. As of January 29, 2013:
http://www.af.mil/information/factsheets/factsheet.asp?id=92

———, "C-5 A/B/C Galaxy and C-5M Super Galaxy," fact sheet, August 16, 2012. As of January 29, 2013:
http://www.af.mil/information/factsheets/factsheet.asp?id=84

U.S. Army and U.S. Marine Corps, *Gaining and Maintaining Access: An Army–Marine Corps Concept*, version 1.0, March 2012.

U.S. Department of Defense, *Joint Operational Access Concept (JOAC)*, version 1.0, Washington, D.C., January 17, 2012. As of January 29, 2013:
http://www.defense.gov/pubs/pdfs/JOAC_Jan%202012_Signed.pdf

"USS *Cole* Bombing: Suicide Attack," *GlobalSecurity.org*, last updated October 11, 2006. As of January 29, 2013:
http://www.globalsecurity.org/security/profiles/uss_cole_bombing.htm

"The WS MRL Family," online discussion thread, *World Affairs Board*, last updated January 28, 2011. As of January 29, 2013:
http://www.worldaffairsboard.com/ground-warfare/58474-ws-mrl-farmily.html

Vick, Alan J., *Snakes in the Eagle's Nest: A History of Ground Attacks on Air Bases*, Santa Monica, Calif.: RAND Corporation, MR-553-AF, 1995. As of January 29, 2013:
http://www.rand.org/pubs/monograph_reports/MR553.html